STEM Careers

CIVIL ENGINEER

Tammy Gagne and
Heather Kissock

AV2
www.av2books.com

Step 1
Go to **www.av2books.com**

Step 2
Enter this unique code

POILG6AS1

Step 3
Explore your interactive eBook!

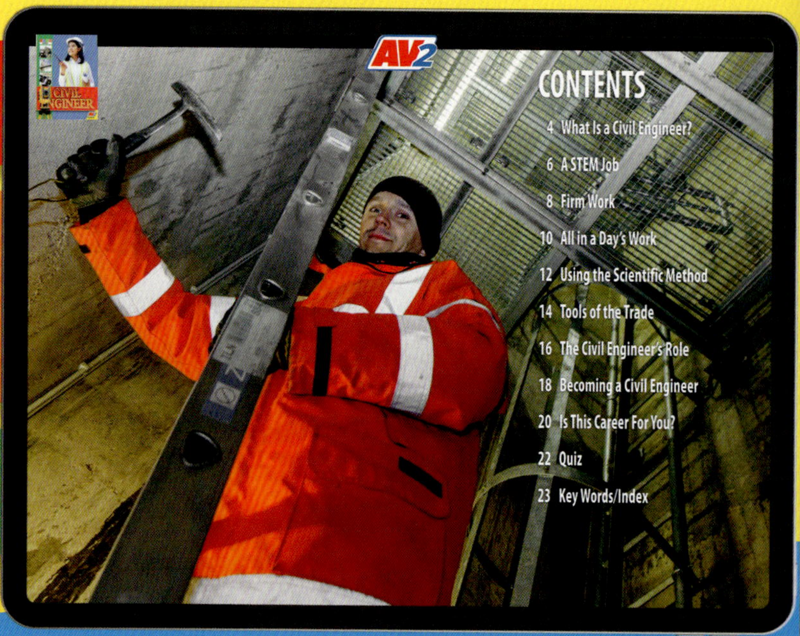

AV2 is optimized for use on any device

Your interactive eBook comes with...

Contents
Browse a live contents page to easily navigate through resources

Audio
Listen to sections of the book read aloud

Videos
Watch informative video clips

Weblinks
Gain additional information for research

Try This!
Complete activities and hands-on experiments

Key Words
Study vocabulary, and complete a matching word activity

Quizzes
Test your knowledge

Slideshows
View images and captions

... and much, much more!

STEM Careers

CIVIL ENGINEER

Contents

What Is a Civil Engineer?

A civil engineer is a person involved in the creation of structures. Civil engineering can include the development of buildings, roads, and bridges. A civil engineer may design these structures or oversee their construction. Civil engineers are also responsible for upgrading these structures as they age. The word *civil* means "relating to citizens." Civil engineers focus on making structures safe and **functional** for those who use them.

Civil engineering may sound like a modern job, but it is actually one of the world's oldest professions. Civil engineers have existed since people began building primitive shelters and laying tree trunks over rivers to make bridges. The structures civil engineers have built over time have helped advance **civilization** to what it is today.

Approximately **311,000** civil engineers work in the **United States**.

The **average salary** of a civil engineer in the United States is **$94,360**.

New York's **Rensselaer Polytechnic Institute** awarded the country's **first** civil engineering **degree in 1835**.

Civil engineers are an important part of a construction team. They are heavily involved in project planning and must monitor work on an ongoing basis to ensure that it is meeting expectations.

A STEM Job

Civil engineers work in the field of STEM. STEM stands for Science, Technology, Engineering, and Mathematics, or math. STEM is a developing field. Jobs in STEM change as new tools and techniques are created. As technology is always advancing, people with degrees in STEM fields are in high demand.

STEM professions use science, technology, engineering, and math to solve problems. Some STEM professions rely on only one or two of these subject areas. Civil engineering uses all four.

Where Civil Engineers Work

Civil engineers can be found across the United States. However, the state offering the most opportunities in this field is California. This is largely because the state is also home to the most technology companies in the nation.

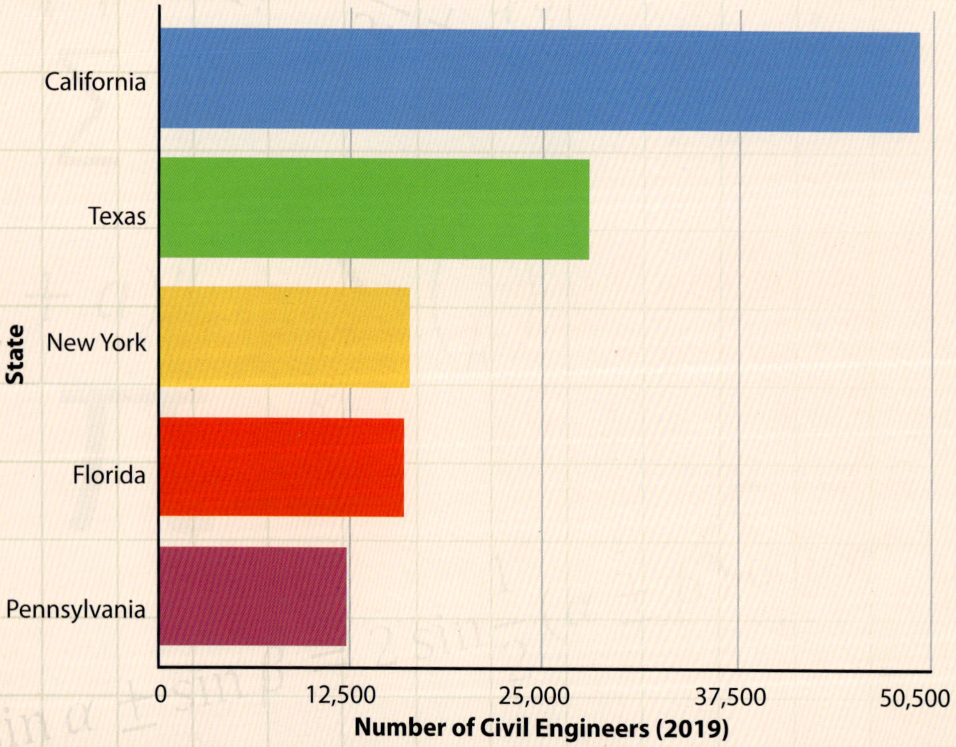

Civil engineers are problem solvers. Some of them use science to help design buildings that can withstand earthquakes. Other civil engineers use technology to build dams that produce electricity. Understanding engineering is important for building bridges that people and cars cross every day. Math is also an essential skill in this profession. Civil engineers use math formulas to measure the safety of the materials used in all types of projects.

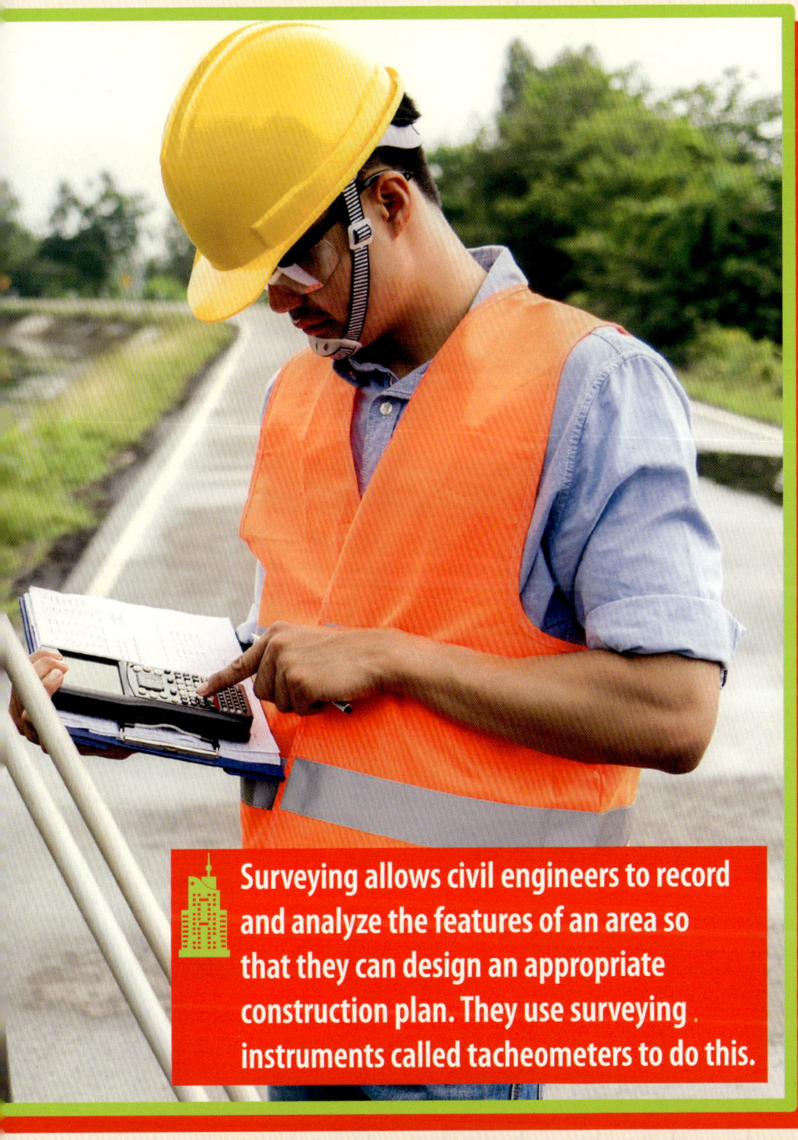

Surveying allows civil engineers to record and analyze the features of an area so that they can design an appropriate construction plan. They use surveying instruments called tacheometers to do this.

STEM

Civil engineers have a **low unemployment rate** of just **0.9 percent.**

In 2020, *U.S. News & World Report* ranked **civil engineering** as one of its **Top 30 Best STEM Jobs.**

More than **13,000 students** graduate with civil engineering degrees each year in the **United States.**

Firm Work

Civil engineers work in numerous fields. In the United States, many find jobs with the government. These civil engineers design, build, operate, and maintain construction projects. They may work on the development of airports, sewage treatment plants, or other projects. They might also work for private companies that the government hires for these projects. Some civil engineers put their skills to use working for transportation businesses, **patent** attorneys, or consulting firms.

Engineering Firms in the United States

The United States has many firms that focus solely on engineering work. Civil engineers are usually key employees in these firms.

HDR. Inc.
Omaha, Nebraska

AECOM

AECOM Technology
Los Angeles, California

Many civil engineers work for architecture firms. These companies design structures such as schools, stadiums, and even outdoor spaces such as large parks. Civil engineers work closely with the architects on these projects to make sure that every part of the structure is safe. They also make sure that the area's zoning laws are followed. Zoning laws are the rules a city or town has for the size and placement of buildings.

Bayside Engineering, Inc.
Woburn, Massachusetts

Bechtel
Reston, Virginia

A civil engineer may be called upon to check the integrity of an older structure, such as a bridge. By performing a series of tests, the engineer will be able to tell if the structure needs repairs.

All in a Day's Work

Civil engineers can perform a range of tasks in a single day. They may create an estimate for a new project. This requires them to research and write a detailed prediction of what a job will cost. Some civil engineers may head outside the office to take soil tests at a job site. They will then have to analyze the results and prepare a report when they return. Civil engineers often attend meetings throughout the day as well.

A civil engineer must be **professional** at all times. He or she must be able to develop a realistic plan and see it through to completion. The engineer must also communicate this plan to the client and to the people working on the project. Most civil engineering projects have strict deadlines. The civil engineer needs to address problems as they occur and keep the job moving forward.

Sample Daily Schedule

A civil engineer has to see the big picture while focusing on the many small details that can make or break a project.

9:00 AM

A civil engineer arrives at work to find a request from a nearby city's planning board. It asks him to do a **feasibility study** for replacing a bridge in the heart of the city. After adding the study to his calendar, he replies that he will present the information to the board at its next meeting.

10:00 AM

The civil engineer receives a message that another town's planning board has approved a new fire station. The civil engineer begins preparing the **contract** so that construction can begin.

12:00 PM

The engineer heads to a construction site. The crew has nearly completed work on a new intersection. It is scheduled to re-open in one week. The civil engineer checks to make sure that the intersection will be done on time.

1:00 PM

The civil engineer goes across town to inspect a new apartment building. Electricians finished the wiring, and the contractor needs paperwork stating that everything was done properly before he can move on to the next phase of construction.

2:00 PM

An email arrives from a **drafter** at the engineer's firm. This person has planned the layout for a new **subdivision** and needs some input. The engineer goes over the plan and sends it back to the drafter with suggestions for changes.

4:00 PM

The civil engineer learns that a large department store may be coming to the area. He starts organizing a **proposal** to send to the store's owner. He wants to convince the owner that his engineering company can handle the project, from planning to completion.

Using the Scientific Method

Scientists use the scientific method to solve problems. The scientific method is a series of steps. It begins with a question scientists want to answer. First, the scientists find out if someone else has asked the same question. If so, they review that scientist's results and gather information for their work. Next, they make a hypothesis. This is what they think the answer to their question will be. The scientists then test the hypothesis with an experiment. After the experiment, they form a conclusion. This means they study the results and decide if the question was answered. Finally, the scientists share the results with the public and other scientists.

1. Ask a Question

Can eggshells support the weight of a pile of books without breaking?

2. Construct a Hypothesis

When an eggshell is broken in half, each part forms an arch. Arches are said to be one of the strongest structures used in building. If organized properly, eggshells should be able to hold the weight of the books.

3. Test with an Experiment

Gather six eggs, cellophane tape, scissors, and six hardcover books. Break off the smallest end of each egg. Empty the eggs' contents into a bowl and put it away for later use. Take six strips of tape and wind them around the center of each shell. Cut through the center of the tape to create six dome-shaped shells. Lay the domes, with the flat side down, on a table to form a rectangle. Now, lay books one by one on the eggshells. See if you can place all of the books on the shells without having them break.

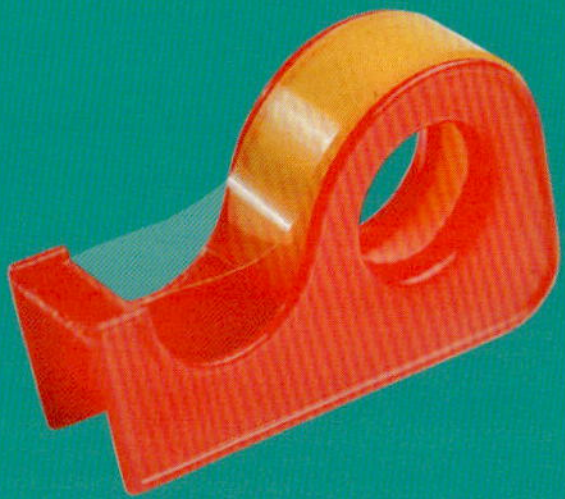

4. Did the Test Work?

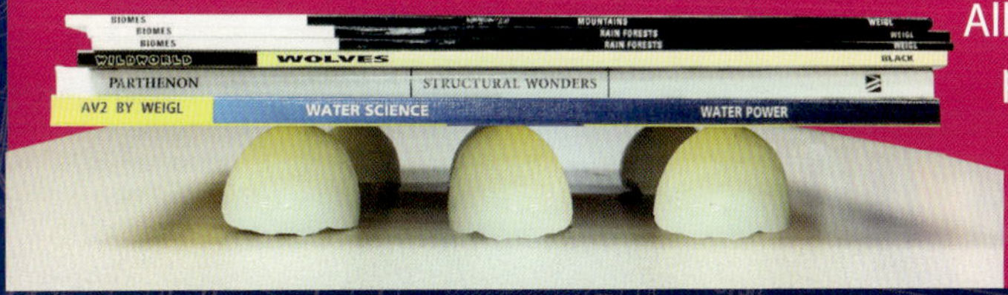

All of the books were placed on the eggshells without any shells cracking or breaking.

5. Draw Conclusions

Arches can be used to support heavy weights. They can also be made with few materials. This makes arches strong and cost-efficient structures to use in construction projects.

Tools of the Trade

Civil engineers use a variety of tools. Some are simple devices that many people use. Others are more complex tools used primarily by engineers and scientists.

Distance Meter

Taking measurements is something civil engineers do almost daily. To get very accurate measurements, they often use distance meters. These tools use **lasers** to take various measurements, such as length, surface area, and volume.

Traffic Counter

Civil engineers use traffic counters to find out how many vehicles pass through certain areas. These long tubes are placed across a road. Each time a vehicle passes over one, sensors in the tube send a signal to a computer, which records the data. Civil engineers can then use this information to decide what can be done to improve traffic flow in the area.

Computer-Aided Design (CAD)

Civil engineers use computer-aided design (CAD) software to help them plan the structures they want to build. The software allows engineers to design a structure on a computer. They can then look at a model of the structure and view it from various angles. This allows them to see how the structure will function before building it.

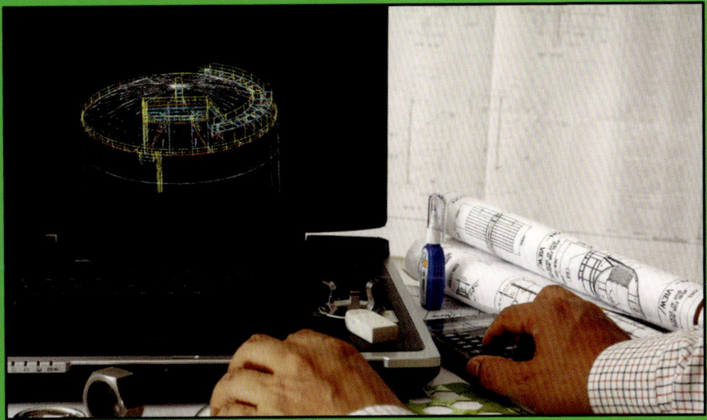

Tablet

Another simple yet essential tool for civil engineers is a tablet. Civil engineers make constant visits to job sites and are often in meetings. They need to make notes and record data while there. A tablet allows them to do all of this and is also easy to carry from one place to another.

Counting Cars

Early traffic counters only recorded how many vehicles traveled over them during a given period. Today, these instruments can collect other types of information as well.

Then

The first counters were called traffic recorders. They were made up of a strip that laid across a road and a counting unit, which was powered by a 6-volt battery. Vehicles were counted as they went over the strip. A report was printed every hour.

New

While traditional traffic counters are still used, advances in technology have allowed them to perform even more functions. For example, some traffic counters can now record the speeds of passing vehicles. Others can be installed in the road itself. They detect the presence of a vehicle at an intersection and help to time the traffic lights for the amount of traffic.

The Civil Engineer's Role

Designing and building safe, reliable structures has always been at the heart of civil engineering. Civil engineers oversee large construction projects. In the past, these projects took a long time to plan and even longer to build. This is because the engineers were limited by the technology of their individual eras. Designing projects with hand drawings took time. Construction crews also had to do all of their work on site, sometimes in harsh weather.

Many of today's apartments, schools, and office buildings are built using modular construction methods.

Today, engineers and construction crews can perform many tasks more quickly and easily. The use of computers makes designing a much faster process. Construction itself can be done more quickly as well. Many of today's structures are built in **modular** sections. This means that they are produced in one place and assembled in another. Entire sections are now often built on **assembly lines**. This speeds up the process and makes construction less expensive.

Modular Construction

Modular construction has many advantages over traditional building methods. One is its speed. Building a modular structure can reduce construction time by about half.

Between **60** and **90 percent** of a modular construction project is completed **before transport**.

Most modular construction takes about **3 to 5 months** to complete.

Some **modular buildings** have been completed in as little as **10 days**.

Becoming a Civil Engineer

Students who want to become civil engineers demonstrate an early interest in construction and problem solving. As children, most civil engineers favored building blocks at playtime. Most of them could quickly visualize what they wanted to create long before placing the first two pieces together. They also had enough patience and persistence to follow the project through to completion.

Future civil engineers are usually excellent students in general, with a deep interest in math and science. In high school, their studies often include advanced math classes such as calculus and trigonometry. Many also belong to clubs that build model trains or airplanes.

Civil Engineer Salaries

A civil engineer's salary is based on his or her experience and training. A salary will increase as the engineer works on more projects and gains more experience.

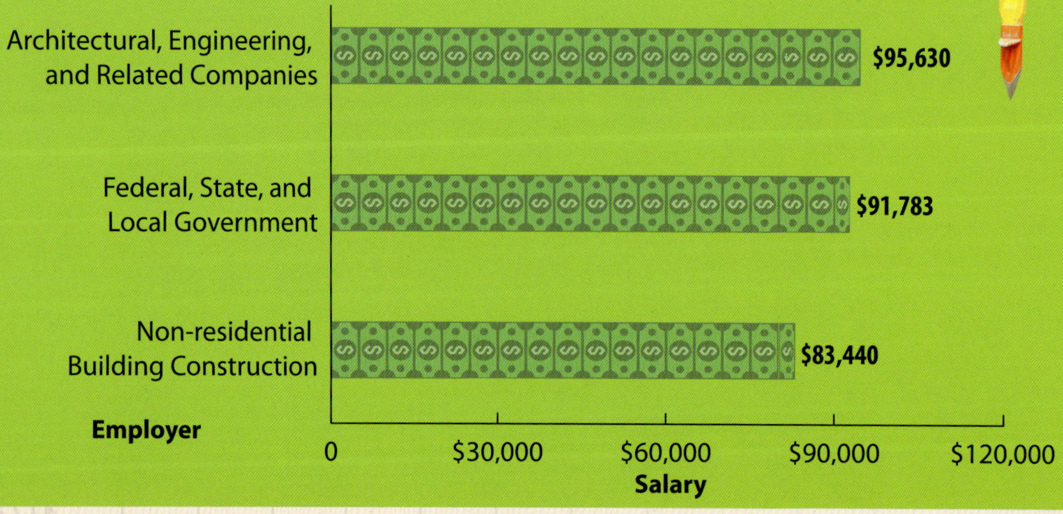

Civil engineers are often responsible for inspecting community construction projects, such as the replacement of water or gas pipes.

Is This Career For You?

Civil engineering students spend at least four years in college. To succeed in this field, students must show that they have solid organizational skills and the ability to make good decisions. They must also be able to communicate difficult concepts in a clear and concise manner.

✓ Education

Most jobs in this field require at least a **bachelor's degree** in civil engineering. Engineers in leadership roles tend to have a **master's degree** in a specialized area such as structural or transportation engineering.

✓ Training

Most engineering degrees require that students complete design projects. These hands-on activities allow them to apply the skills they are learning. Students may also be encouraged to apply for **internships**. These opportunities help them gain practical experience by showing them the common problems that can arise and how to solve them.

✓ Application

Before a civil engineer can work for the public, he or she must become a licensed professional engineer (PE). Licensing requirements vary by state, but most require the civil engineer to have experience in the field. A candidate must also pass exams that demonstrate his or her engineering knowledge.

Career Connections

Plan your civil engineering career with this activity. Follow the instructions in the steps below to complete the process of becoming a civil engineer.

A Visit the civil engineering department of a construction company or architecture firm. Talk to civil engineers. Ask them what they like about their work. Find out where they went to college.

B Decide if this is a job you want to do. Do you want to spend many hours planning large-scale projects and following them through to completion?

C Think about the skills you need to be a civil engineer. People who work in this field must be good at performing complex math, running advanced computer programs, and communicating with others.

D Visit an educational fair at a college that teaches engineering classes. Find out which high school classes will prepare you to pursue a civil engineering degree.

1. Think about how you can get work experience while at college. Meet with your professors. Ask them about summer jobs.

2. Attend career fairs at your college. Talk to people who hire civil engineers. Ask for advice on how to apply for a job. Find out where their companies post job information.

3. Follow civil engineering job sites on the internet. Apply for jobs with companies that have posted openings. Look for internships with state and local governments.

4. Be well prepared for a job interview. Research the organization. Know what they do. Make a list of questions about the job being offered. Employers will ask questions about your education. Make sure to tell them about your work experience.

Quiz

1. What does the word "civil" mean?
2. What does "STEM" stand for?
3. What does a feasibility study explore?
4. What is a hypothesis?
5. What types of measurements can distance meters take?
6. What type of construction involves creating entire sections of buildings on assembly lines?
7. What kinds of toys did many civil engineers play with as children?
8. Calculus and trigonometry are advanced forms of which subject?
9. What level of education do most civil engineers need to find work in the field?
10. What does CAD stand for?

Answers

1. Relating to citizens
2. Science, Technology, Engineering, Mathematics
3. Whether a plan will work or not
4. A guess about what will happen in an experiment
5. Length, surface area, and volume
6. Modular construction
7. Building blocks
8. Mathematics
9. Bachelor's degree
10. Computer-aided design

Key Words

assembly lines: arrangements of workers and machines in factories, where each worker deals with only one part of a product

bachelor's degree: the first level of a college program of study

civilization: the society, culture, and way of life of a particular area

contract: an agreement that can be enforced by law

drafter: a person who draws plans and sketches for machinery or structures

feasibility study: an assessment of whether a plan will work or not

functional: practical and useful

internships: positions a student takes to gain experience, often without pay

lasers: devices that create an intense beam of light

master's degree: the second-highest level of a college program of study

modular: involving a set of standardized parts

patent: the granting of a property right to an inventor

professional: exhibiting a courteous, conscientious, and businesslike manner in the workplace

proposal: a written plan created for the consideration of others

subdivision: an area of land containing many homes built at about the same time

Index

Published by AV2 by Weigl
14 Penn Plaza, 9th Floor
New York, NY 10122
Website: www.av2books.com

Library of Congress Cataloging-in-Publication Data
Names: Gagne, Tammy, author. | Kissock, Heather, author.
Title: Civil engineer / Tammy Gagne.
Description: New York, NY : AV2, [2021] | Series: STEM careers | Includes index. | Audience: Ages 9-12. | Audience: Grades 4-6. |
Identifiers: LCCN 2019042816 (print) | LCCN 2019042817 (ebook) | ISBN 9781791116804 (library binding) | ISBN 9781791116811 (paperback) | ISBN 9781791116828 | ISBN 9781791116835
Subjects: LCSH: Civil engineering--Juvenile literature.
Classification: LCC TA149 .G34 2021 (print) | LCC TA149 (ebook) | DDC 624.023--dc23
LC record available at https://lccn.loc.gov/2019042816
LC ebook record available at https://lccn.loc.gov/2019042817

Printed in Guangzhou, China
1 2 3 4 5 6 7 8 9 0 24 23 22 21 20

052020
101119

Editor: Heather Kissock
Designer: Ana María Vidal

Every reasonable effort has been made to trace ownership and to obtain permission to reprint copyright material. The publishers would be pleased to have any errors or omissions brought to their attention so that they may be corrected in subsequent printings.

The publisher acknowledges Getty Images, Alamy, iStock, and Shutterstock as its primary image suppliers for this title.